Design, Construction and Analysis of 1 KVA Inverter Powered by Solar Panel via a 12V Industrial Battery

Idris Abidemi Akinloye
Olatona G.I.
Ogunbiyi D. M.
Sulaiman F. A.
Adegbola F.

ELIVA PRESS

ELIVA PRESS

Idris Abidemi Akinloye
Olatona G.I.
Ogunbiyi D. M.
Sulaiman F. A.
Adegbola F.

The need for an uninterrupted power supply cannot be over emphasized. Solar energy is one of the cleanest and renewable sources of energy available. However, it comes in form of direct current while most of the electric devices are designed to operate on alternating current. Hence this work concerns the design, construction and analysis of 1000 VA inverter powered by a solar panel via a 12 V industrial battery. The solar panel is made to charge the battery using a solar charge controller as a regulator. The 12 V DC from the battery is converted to 12 V AC at a frequency of 50Hz using integrated circuits, semiconductors and a power MOSFETs as a switching component. This voltage is then stepped up to a 220 V AC via the windings of a transformer. The circuit is found to be stable when within ±3% of the rated power.

Published: Eliva Press SRL
Address: MD-2060, bd.Cuza-Voda, 1/4, of. 21 Chişinău, Republica Moldova
Email: info@elivapress.com
Website: www.elivapress.com

ISBN: 978-1-952751-34-9

DESIGN, CONSTRUCTION AND ANALYSIS OF 1 KVA INVERTER POWERED BY SOLAR PANEL VIA A 12V INDUSTRIAL BATTERY

BY

AKINLOYE IDRIS ABIDEMI

A PROJECT SUBMITTED TO

THE DEPARTMENT OF PHYSICS,

FACULTY OF BASIC AND APPLIED SCIENCES.

COLLEGE OF SCIENCE, ENGINEERING AND TECHNOLOGY.

1

DEDICATION

This project is dedicated to Almighty Allah, the lord of all that exist.

ACKNOWLEDGEMENT

I wish to express my profound gratitude to Almighty Allah who has kept me by His grace, for bringing me this far in my academic pursuit and for the successful completion of this work. I am greatly indebted to my family and my guidance Alh. Nureen O. AKINLOYE and Mrs. Sherifah. Akinloye for their parental, financial support and their words of encouragement. I pray that may you reap the fruit of your labor.

I sincerely appreciate the indispensable role played by my supervisor Dr. Gbadebo I. Olatona; not only for his diligent and painstaking guidance through this work but also for many useful suggestions and constructive criticism as well as thorough reading through the completed manuscript. I will forever be grateful and indebted to him.

I am grateful to the Head of the department Dr. Olusegun O. Alabi and all lecturers (Dr. Gbadebo I. Olatona, Dr M. A. Fakunle, Dr. O. A. Ojo, Dr. J. T. Adeleke and Dr. F.O. Oladejo) of the Department of Physics, Osun State University, Osogbo, for their guidance and encouragement throughout the duration of my program and for positive criticisms during my seminar presentations.

I also wish to express my gratitude to all others who contributed to the success of this work and also to the success of my undergraduate program.

My sincerely appreciation also goes to the best clique Shaykh Shuaib, Shaykh yuusuf, Eng. S.O. Oladunjoye and CO. of LADWOL NIG. LTD., My hostel community members (The OLADUA family, Ummu Abdullah and Imam Taofeeq) and my project mates(Sulaimon F.,Ogunbiyi D. and Agboola F.) I associated with for their love and support throughout the undergraduate program; I say a very big thanks.

3

ABSTRACT

The need for an uninterrupted power supply cannot be over emphasized. Solar energy is one of the cleanest and renewable sources of energy available. However it comes in form of direct current while most of the electric devices are designed to operate on alternating current. Hence this work concerns the design, construction and analysis of 1000 VA inverter powered by a solar panel via a 12 V industrial battery. The solar panel is made to charge the battery using a solar charge controller as a regulator. The 12 V DC from the battery is converted to 12 V AC at a frequency of 50Hz using integrated circuits, semiconductors and a power MOSFETs as a switching component. This voltage is then stepped up to a 220 V AC via the windings of a transformer. The circuit is found to be stable when within ±3% of the rated power.

KEY WORDS: Solar energy, Inverter, Transformer, MOSFET, Current, Conversion.

CHAPTER ONE

INTRODUCTION

1.1   Background Study

Renewable energy is a form of energy that is derived from renewable resources, which are naturally replenished on a human timescale, such as sunlight, wind, rain, tides, waves, and geothermal heat.[1] Renewable energy often provides energy in four important areas: electricity generation, air and water heating/cooling, transportation, and rural (off-grid) energy services. Renewable energy is derived from natural processes that are replenished constantly. In its various forms, it derives directly from the sun, or from heat generated deep within the earth. Included in the definition are electricity and heat generated from solar, wind, ocean, hydropower, biomass, geothermal resources, bio-fuels and hydrogen derived from renewable resources [2].

Renewable energy sources exist over wide geographical areas, in contrast to other energy sources, which are concentrated in a limited number of countries. Rapid deployment of renewable energy and energy efficiency is resulting in significant energy security, climate change-mitigation, and economic benefits. The results of a recent review of the literature concluded that as greenhouse gas (GHG) emitters begin to be held liable for damages resulting from GHG emissions resulting in climate change, a high value for liability mitigation would provide powerful incentives for deployment of renewable energy technologies [3].

National renewable energy markets are projected to continue to grow strongly in the coming decade and beyond. At least two countries, Iceland and Norway, generate all their electricity using renewable energy already, and many other countries have the set a goal to reach 100% renewable energy in the future. At least 47 nations around the world already have over 50 percent of electricity from renewable resources [4]. Renewable energy resources exist over

5

wide geographical areas, in contrast to fossil fuels, which are concentrated in a limited number of countries. Rapid deployment of renewable energy and energy efficiency technologies is resulting in significant energy security, climate change mitigation, and economic benefits. In international public opinion surveys there is strong support for promoting renewable sources such as solar power and wind power. While many renewable energy projects are large-scale, renewable technologies are also suited to rural and remote areas and developing countries, where energy is often crucial in human development. Most of renewable energy technologies provide electricity, renewable energy deployment is often applied in conjunction with further electrification, which has several benefits: electricity can be converted to heat (where necessary generating higher temperatures than fossil fuels), can be converted into mechanical energy with high efficiency, and is clean at the point of consumption. In addition, electrification with renewable energy is more efficient and therefore leads to significant reductions in primary energy requirements, because most renewable energy technologies do not need a thermodynamic cycle with high losses.

The History of Renewable Energy Achievements

A historical review, even a brief one, is not always an easy task. There are many uncertainties and few, very scarce documents, not easily accessible. Nevertheless historical references and in general historical events that promoted science and technology are of great importance to mankind [5]. Old knowledge may help to adapt, sophisticate ideas and philosophical thoughts about physical phenomena, to the knowledge of our time, and find ways for applying these ideas using the modern technological achievements. It is obvious that many good ideas found no application due to technological limitations at the time they were expressed. These energies are applied directly to the desalination methods and/or they are potential sources for large scale desalination plants. The historic review starts with early philosophical thoughts about the natural forces and physical phenomena, and continues with the progress made by the scientific explanations given by ancient philosophers, until the

6

sophisticated achievements and applications of our times. Emphasis is given to those achievements whose impact on civilization and technological progress appears more significant. Units, if they exist, are given in their original symbols, or expressions with the IU System in parentheses. Of all natural phenomena, these related to the renewable energies, such as intensive solar radiation, destructive phenomena, such as floods, and volcano eruptions, impressed early mankind, who was unable to explain them - and thus in many cases believed them as Gods. Little by little the philosophers, thinking deeply, observed these natural phenomena and discovered that natural forces could be trapped and utilized for the benefit of their families or their nations.

The ability of biomass and biofuels to contribute to a reduction in CO_2 emissions is limited because both biomass and biofuels emit large amounts of air pollution when burned and in some cases compete with food supply. Furthermore, biomass and biofuels consume large amounts of water. Other renewable sources such as wind power, photovoltaic's, and hydroelectricity have the advantage of being able to conserve water, lower pollution and reduce CO_2 emissions.

TYPES OF RENEWABLE ENERGY

The following are the types of renewable energy in the world

1. Hydropower

2. Wind power

3. Solar power

4. Geothermal Power

5. Bio power

HYDROPOWER RENEWABLE ENERGY:

This form uses the gravitational potential of elevated water that was lifted from the oceans by sunlight. It is not strictly speaking renewable since all reservoirs eventually fill up and require very expensive excavation to become useful again. At this time, most of the available locations for hydroelectric dams are already used in the developed world.

The most common method of hydropower generation involves construction of dams on rivers and releasing water from the reservoir to drive turbines. Pumped-storage type plants represent another method of hydroelectricity generation.

Hydropower is conceivably regarded as the major source of electric power generation and supply in Nigeria because the country is endowed with large rivers, waterfalls and dams. Only large hydropower technology is the prominent commercial technology in the electricity supply mix of the country. Due to economy of scale, large hydropower technology takes the lion share of the entire commercial renewable energy resources for electricity generation under any CO_2 emission constraints. Unlike fossil fuel, hydropower is renewable and can supply uninterrupted fuel, except for the question of water levels. The total potential of hydropower in Nigeria is about 14,750 MW. However, only1930 MW, approximately 14%, of that is currently being generated at Shiroro, Kanji and Jebba representing about 30% of gross installed grid-connected electricity generation capacity of Nigeria [6].

WIND POWER RENEWABLE ENERGY:

Wind turbines work by pooling the mechanical energy of wind and transforming it into electricity. Wind turns the blades, which spin a shaft, which then connects to a generator and produces electricity. Wind turbines can be built on land or offshore in large bodies of water like oceans and lakes. Based on data from EWEA (European Wind Energy Association),in 2010 there were 70,488 onshore wind turbines and 1,132 offshore turbines across the EU. With technological progress turbines are becoming bigger and more efficient: the same

8

amount of energy can be generated with fewer turbines. An average size onshore turbine manufactured in 2014 can power annually more than 1,500 average EU households. An average offshore wind turbine of 3.6 MW can power more than 3,312 average EU households.

The movement of the atmosphere is driven by differences of temperature at the Earth's surface due to varying temperatures of the Earth's surface when lit by sunlight. Wind energy can be used to pump water or generate electricity, but requires extensive areal coverage to produce significant amounts of energy. Wind is the second most widely used renewable source, as global wind power installed capacity exceeded 283GW in 2013. The annual growth rate of cumulative wind power capacity has averaged 25% during last five years making wind the fastest growing renewable power source, a trend projected to continue in the future [7].

SOLAR POWER RENEWABLE ENERGY:

This form of energy relies on the nuclear fusion power from the core of the Sun. This energy can be collected and converted in a few different ways. The range is from solar water heating with solar collectors or attic cooling with solar attic fans for domestic use to the complex technologies of direct conversion of sunlight to electrical energy using mirrors and boilers or photovoltaic cells. Unfortunately these are currently insufficient to fully power our modern society. Global installed capacity of solar power is more than 100GW, which makes it the third biggest renewable power source, with photovoltaic (PV) technology the dominant source. The use of concentrating solar power (CSP) technology is also on the rise and global installed capacity stood at 2.5GW at the beginning of 2013.

GEOTHERMAL POWER RENEWABLE ENERGY:

Energy left over from the original accretion of the planet and augmented by heat from radioactive decay seeps out slowly everywhere, everyday. In certain areas the geothermal gradient (increase in temperature with depth) is high enough to exploit to generate electricity. This possibility is limited to a few locations on Earth and many technical problems exist that limit its utility. Another form of geothermal energy is Earth energy, a result of the heat storage in the Earth's surface. Soil everywhere tends to stay at a relatively constant temperature, the yearly average, and can be used with heat pumps to heat a building in winter and cool a building in summer. This form of energy can lessen the need for other power to maintain comfortable temperatures in buildings, but cannot be used to produce electricity. Installed power production capacity from geothermal sources exceeded 11.7GW as of 2013 making it the fifth biggest renewable source for electricity generation. [8].

BIOPOWER RENEWABLE ENERGY:

This is the term for energy from plants. Energy in this form is very commonly used throughout the world. Unfortunately the most popular is the burning of trees for cooking and warmth. This process releases copious amounts of carbon dioxide gases into the atmosphere and is a major contributor to unhealthy air in many areas. Some of the more modern forms of biomass energy are methane generation and production of alcohol for automobile fuel and fueling electric power plants.

Bio-power is the fourth biggest renewable power source after hydro, wind and solar. The world's net electricity production capacity from bio-mass currently exceeds 83GW, while global bio-power generation increased from 313TWh in 2010 to more than 350TWh in 2013[9]. Modern biomass, especially bio-fuels and wood pellets, are increasingly being used for heat and power generation, alongside traditional biomass sources such as agricultural bioelectricity.

1.2 AIMS AND OBJECTIVE OF THE PROJECT

One of the significant problems years' back which has been yearning for long lasting solution is the epileptic supply of electricity. Most especially in Nigeria, it has become rampant and has made life boring for individuals, organizations sector and industries. Some researchers have proposed the use of stand by generators with attendant noise and partial daily maintenance in cost of fuelling. Those that could not avoid generator for their aims and waiting for appropriation of the electricity supply thereby continue in darkness in case of power failure.

A renewable technological advancement has become as way out to this prominent problem through the design and construction of a hybrid inverter. It efficiently serves as a means of generating electrical power supply without any interruptions in any way by its inbuilt mechanism and automatic regulator (operator).

The main objective of this project is to Design and Construct an inverter which can be powered from the source of 12V battery which includes a regulator circuit for its battery to charge it through solar source to give an output of both 220V for AC appliances and 12V or DC appliances. This Inverter will be able to power specified electronics loads in wattage within the range of 1 to 800 watts such as lightning point, computers, fridge, and television. The aim of the study is to evaluate and analyze the performance of a branded and locally constructed Modified 1KVA Sine Wave Solar Power Inverter for domestic electric power supply.

BASIC PRINCIPLE OF THE PROJECT

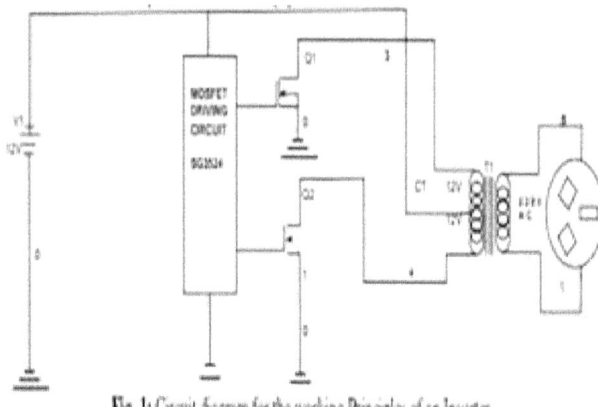

Fig. 1: Circuit diagram for the working Principles of an Inverter

The basic working principles of an inverter can be explained with the aid of Fig1 above. The inverter operates by performing two functions. First, it converts DC voltage (12V) from the battery to AC voltage (220V) using a pair of powerful MOSFETs (Q1 and Q2) serving as switches. Then, it steps up the resulting AC into equivalent mains voltage with corresponding frequency and phase, using appropriate step up transformer.

1.3 SIGNIFICANCE OF THE PROJECT

The use of solar inverter is to generate energy provide many benefits to the community which includes;

1. Solar power inverter wills priorities charging from solar panels, enabling your batteries to charge via the PV panels even when public power supply is on, leading also to savings on your power bills. Reduction in the cost associated with generating electricity.

2. Solar inverter has always helped in reducing global warming and green house effect. Also use of solar inverter helps in saving money that would have used for buying fuel for conventional generator.

3. Solar inverters will allow you charging to solar panels or power grid depending the battery level. Some solar inverters are even intelligent enough just to take just as much deficit current from the grid as is required.

12

4. A solar inverter helps in converting the Direct current in batteries into alternative current. This helps people who use limited amount of electricity.

5. Solar inverters are the best way and they are better than the normal electric ones. Also their maintenance does not cost much money.

6. It also helps the country in areas of technical advancement by carryout more skill importance project work that will encourage acquisition of more skill for sustenance of technological development.

7. It can also be used in our home and labs to power electronics appliances such as radio-set, computer etc. High quality and low cost; for powering offices which has limited electronics load like light point, fan, charging port etc.

8. Inverter is easy to operate in all conditions.

9. It produces a stable voltage supply without fluctuating.

10. Inverter are much quieter compared to conventional ones i.e. the outdoor unit usually makes far less sound as the unit is operating at a reduced rate.

1.4 LIMITATION OF THE PROJECT

Limitations are as follows;

1. **Operating Power**: The power of this device is 1000VA that means any load above 900 Watt should not be powered by this device. It's easy to damage the machine if the user is not familiar with the operation; the user needs to strictly follow the manufacturers' guide.

2. **Over Charge**: The intensity of the Sun varies throughout the day. This creates an over-charging problem if the panels are connected to the battery directly, so for this reason a charge controller must be used to offer protection from high voltage and current from the panels.

3. **Operating Frequency**: The inverter shall be capable of operating at the nominal frequency (50Hz).

4. **Ventilation**: The inverter isolation transformer shall be designed for convection cooling. If a fan cooling is required for the MOSFET used, then iron and a good heat absorption casing will be used.

13

5. **Response Time**: The inverter shall respond to any line voltage variation in 1/2 cycle while operating linear or non-linear loads, with a load power factor of 0.80 of unity. Peak detection of the voltage sine wave shall not be permitted to avoid inaccurate tap switching due to input voltage distortion.

6. **Access Requirements**: The inverter shall have removable panels for ease of maintenance and repair.

7. **Metering**: An input meter is provided to display line voltages of the battery during charging.

1.5 APPLICATIONS OF INVERTER

1. In power distribution networks AC/DC converters may be used to exchange power between utility frequency 50 Hz and 60 Hz power grids.

2. DC converters (inverters) are used primarily in UPS or renewable energy systems or emergency lighting systems. When mains power is available, it will charge the DC battery. If the mains fail, an inverter will be used to produce AC electricity at mains voltage from the DC battery.

1.6 BENEFITS OF THE PROJECT

- It serves as a backup for power supply.

- It does not require any special maintenance.

- It supplies a constant voltage that is it does no fluctuate.

1.7 PROBLEM OF THE PROJECT

- Its working efficiency to produce direct current only depends on the intensity of the sunlight

- The solar panels that are used to attract Sunlight require lots of space.

- Maintenance and replacement may require technician effort.

CHAPTER TWO

LITERATURE REVIEW

Inverters are crucial energy conversion components in any renewable energy scheme which converts DC voltage to AC voltage required by most electronics and electrical appliances. The ideal inverter has an efficiency of approximately hundred percent and produces a perfect sinusoidal waveform output. Production of a perfect sinusoidal output waveform will require the inverter to operate as a linear amplifier which reduces the efficiency result. To achieve reasonably high efficiency, inverters replace the temporal variations of a sine wave with waveforms that have square edges. Examples of such waveforms include square waves, modified square wave; sinusoidal pulse width modulation (SPWM) synthesized sine wave, and multilevel waveforms. The process of synthesizing power from DC sources is commonly achieved using two basic methods. In the first method, a high voltage DC source whose average value equals the peak value of the AC voltage to be synthesized is generated from the battery using high frequency DC to DC converters. The AC waveform is then generated from the high voltage source using electronic switches. This approach leads to a compact design because of heavy frequency of 50Hz. The other approach involves the direct generation of AC power from the available DC power using 50Hz step-up transformer and electronic switches. Either of these methods can be used to make square wave, modified sine wave and sinusoidal pulse width modulated sine wave inverters. A circuit block related to inverters is the DC to DC converter which converts input DC power to an output DC power at a different voltage level A class of DC to DC converters known as multi-input converters and have the ability to operate from more than one voltage sources have been developed and analyzed in the literature. These converters are used in renewable energy conversion systems that have multiple sources of energy such as solar and wind energy sources, as exemplified. They adjust their voltage gain in relation to the input DC

15

voltage, as needed to generate the required output voltage using pulse-width modulation techniques.

However, direction of direct current is not reversed. It is direct potential (voltage) at the rectifier terminals that is inverted or reversed. Because potential is reversed with current continuing in the same direction as before, the flow of elecTestwer is also reversed or transferred from the DC system to the AC system. An inverter is a motor control that adjusts the speed of an AC induction motor. It does this by varying the frequency of the AC power to the motor. An inverter also adjusts the voltage to the motor. This process takes place by using some intricate electronic circuitry that controls six separate power devices. They switch on and off to produce a simulated three phase AC voltage. This switching process is also called inverting DC bus voltage and current into the AC waveforms that are applied to the motor. This led to the name "inverter".

Prince reviews prior art by first examining operation of a single-phase full-wave center-tap rectifier circuit. The DC output of the rectifier circuit includes both passive resistance and reactance. His figures provide ideal waveforms of most circuit variables, including potential and current at AC and DC terminals. On the left side of the Figure 1 is the "rectifier circuit". This figure is identical in topology to Prince's, except that modern symbols for rectifier circuit elements replace his archaic vacuum tube (diode) symbols. With this small change, his figure is identical to any modern single-phase full wave center- tap rectifier circuit.

However the recent advancement in technology and rising cost in the manufacture of modified sine wave inverters has call for a review and analysis of the performance so as to achieve efficiency and also have cost effectiveness in the production for domestic power supply. The process of synthesizing AC power from DC sources is commonly achieved using two basic methods. In the first method, a high voltage DC source whose average value equals the peak value of the AC voltage to be synthesized is generated from the battery using high frequency DC to DC converters. The AC waveform is then generated from the high voltage source using electronic switches. This approach leads to a compact design because heavy 50Hz magnetic are not used. The other approach involves the direct generation of AC power from the available DC power using 50Hz step-up transformer and electronic switches. Either of these methods can be used to make square wave, modified sine wave and sinusoidal pulse width modulated sine wave inverters. A circuit block related to inverters is the DC to Converter which converts input DC power to an output DC power at a different voltage level [9, 10]. A class of DC to DC converters known as multi-input converters and have the ability to operate from more than one voltage sources have been developed and analyzed in the literature. These converters are used in renewable energy conversion systems that have multiple sources of energy such as solar and wind energy sources, as exemplified.

2.1 HISTORY OF INVERTER

Origins of the Inverter

From the late 19th century, through the middle of the 20th century, direct current (DC) to alternating current (AC) power conversion was accomplished using rotary converters, or motor generator sets (M-G sets), In the early 20th century, vacuum tubes and gas filled tubes .began to be used as switches in inverter circuits. The most widely used type of tube was the thyratron.The origins of the electromechanical inverters explained the source of the term inverter.

Early AC to DC converters used an induction AC motor directly connected to a dynamo so that the generator commutation reversed connections at exactly the right moments to produce DC. A later development is the synchronous converter, in which the motor and generator windings are combined into one armature, with slip rings at one and a commutation at the other and only one field frame. The result with either is AC-in, DC-out. With a rotary converter, the DC can be considered to be "mechanically rectified AC". Given the right auxiliary and control equipment, the rotary converter can be "run backwards" thus converting DC to AC. Hence, an inverter is an inverted converter.

David Prince probably coined the term inverter. It is unlikely that any living person can now, establish with certainty that Prince (or anyone else) was the originator of this commonly used engineering term.

However, in 1925 Prince did publish an article in the GE Review titled "The In-verter" Elf. His article contains nearly all important elements required by modern inverters and is the earliest such publication to use that term in the open literature. The idea of using grid control in combination with phase retard to modulate AC power originated with others about four years earlier. However, Prince appears to have been the individual who took Alexanderson's expression "inverted rectification" and created a single English-language word inverter. It conveys the idea of a rectifier except functioning in a inverted mode of operation, hence inverter. What's in a name? That which we call an inverter by any other name would be an inverter.

By 1936, Prince's inverter appeared in literature from all corners of the world, Europe and Japan among them. It was in common use in English technical publications or its equivalent word was used in other languages.

In 1925, Prince defined inverter as the inverse of rectifier. In so doing, he depended upon his audience having a clear mental abstraction of rectifier and built upon their pre-existing concepts. The term rectifier was in common use for more than two decades prior to 1925. It was understood to mean any stationary apparatus or rotating commutator for transforming alternating into direct current. (Rotary converters, later known as synchronous converters, were in use by 1892 to convert AC power into DC power. Rotary converters were manufactured until the 1950s, when germanium diodes became available. When operated to convert DC power to AC power, rotaries were dubbed "inverted rotaries." The distinction between rectifier and converter was sometimes vague, perhaps even arbitrary, but often based on use of static or non-rotating versus rotating parts.) Prince explained that an inverter is used to convert direct current into single or poly phase alternating current. The article explains how "the author took the rectifier circuit and inverted it, turning in direct current at one end and drawing out alternating current at the other." Use of the word inverted conveys the idea of turning something upside down. What was turned upside down? Clearly, he did not mean to invert the rectifier devices) or rectifier circuit; their orientation remains the same. Rather, he meant to invert the function or operation of the rectifier. That is why he said to draw in direct current and push out alternating current, to emphasize a new mode of operation. However, direction of direct current is not reversed. It is direct potential (voltage) at the rectifier terminals that is inverted or reversed. Because potential is reversed with current continuing in the same direction as before, the flow of electric power is also reversed or transferred from the DC system to the AC system. The inverse of rectification was not an obvious extension of prior art. It required several imaginative steps by Prince to bring his readers to comprehend conversion of electric current of one form (direct) to another form (alternating) Among those innovations was grid control of current conduction Prince was not the originator of that idea, but built upon it.

2.2 REVIEW OF HOW TO CHOOSE THE RIGHT INVERTER

Various options are also available. Choosing which one is right inverter from such a long list can be a chore. There is no "best" inverter for all purposes - what might be great for an ambulance would not be suited for an RV. Power output is usually the main factor, but there are many others. There are many factors that go into selecting the best inverter (and options) for your application, especially when you get into the higher power ranges (800 watts or more) [11]. The following basis must be known to choose a right inverter;

WATTS

The poor watt is often misunderstood. Watts are basically just a measure of how much power a device uses, or can supply, when turned on. A watt is a watt - there is no such thing as "watts per hour", or "watts per day". If something uses 100 watts, that is simply the voltage times the amps. If it draws 10 amps at 12 volts, or 1 amp at 120 volts, it is still 120 watts. A watt is defined as one Joule per second, so saying watts per hour is like saying "miles per hour per day".Watt-hoursA watt-hour (or kilowatt hour, kWh) is simply how many watts times how many hours that is used for. This is what most people mean when they say "watts per day". If a light uses 100 watts, and it is on for 9 hours, that is 900 watt-hours. If a microwave uses 1500 watts, and runs for 10 minutes, that is 1/6th of an hour x 1500, or 250 WH.

AMPS

A Coulomb is the charge of 6.24×10^{18} electrons. Therefore, 1 Amp is equal to the charge of 6.24×10^{18} electrons passing a point in a circuit in 1 second.

Amp-Hours

Amp-hours (usually abbreviated as AH) are what most people mean when they say "amps per hour" etc. Amps x time = AH. AH are very important, as it is the main measure of battery capacity. Since most inverters run from batteries, the AH capacity determines how long you can run.

Power Ratings of inverters

Inverters come in size ratings all the way from 50 watts up to 50,000 watts, although units larger than 11,000 watts are very seldom used in household or other PV systems. The first thing you have to know about your inverter is what will be the maximum surge, and for how long. (More about 230 volts pumps etc later). All inverters have a continuous rating and a surge rating. The surge rating is usually specified at so many watts for so many seconds.

This means that the inverter will handle an overload of that many watts for a short period of time. This surge capacity will vary considerably between inverters, and different types of inverters, and even within the same brand. It may range from as little as 20% to as much s 300%. Generally, a 3 to 15-second surge rating is enough to cover 99% of all appliances - the motor in a pump may actually surge for only 1/2 second or so.

General Rules: The inverters with the lowest surge ratings are the high-speed electronic switching type (the most common). These are typically from 25% to 50% maximum overload. This includes most inverters made by Stat power, Exeltech, Power to go, and nearly all the inexpensive inverters in the 50 to 5000-watt range. The highest surge ratings are the transformer based low-frequency switchers. This includes most Xantrex, Magnum, and Outback Power. Surge ratings on these can range up to 300% for short periods. While high-frequency switching allows a much smaller and lighter unit, due to the much smaller transformers used it also reduces the surge or peak capacity.

Pros and Cons: Although the high-frequency switching type doesn't have the surge capacity of the transformer based, they do have some definite advantages. They are much lighter, usually quite a bit smaller, and (especially in the lower power ranges) they are much cheaper. However, if you are going to run something like a submersible well pump, you will need either very high surge capacity or you will need to oversize the inverter above its typical usage, so that even at maximum surge the inverter will not exceed its surge rating. Choice of choosing the right inverter mainly depends on operating power (wattage), voltage (volt) and current (ampere).

2.3 REVIEW OF THE DIFFERENCE BETWEEN SINE WAVE AND MODIFIED SINE WAVE INVERTER

The primary work of an inverter is to convert the DC (Direct Current) power from the battery bank or solar panels to AC (Alternating Current) power needed for most appliances. To do that, it must take the constant DC voltage and change it to a sine wave curve that goes above and below 0 volts. When inverters first came out, the most common way to do this was to make the voltage go straight up and down, creating a blocky signal. This is called modified sine wave, seen in orange in the image below. More advanced modified sine waves make multiple steps, trying to come close to a pure sine wave. DC Voltage, Modified Sine Wave, and Pure Sine Wave Graph [12].

Pure Sine Wave

A pure sine wave inverter is an inverter that power devices or it is suitable for devices that uses motor, a rectifier, DC adaptor and a delicate piece of medical equipment. The major advantage of a sine wave inverter is that all of the equipment which is sold on the market is designed for a sine wave. This guarantees that the equipment will work to its full specifications. Some appliances, such as motors and microwave ovens will only produce full output with sine wave power.

A pure sine wave inverter is more complex than a modified sine wave inverter and as a result, is a more expensive item to purchase but more cost effective and much safer in the long run. Pure sine wave power flows in even, arching waves and is generally needed for newer LED TVs, CFL light bulbs, and inductive loads like brushless motors. These inverters use more sophisticated technology to protect even the most sensitive electronics. They produce power which equals – or is better than – the power in your home. The primary category of devices that run more efficiently with a pure sine wave inverter is electronics that use AC motors, like refrigerators, compressors, and microwave ovens.

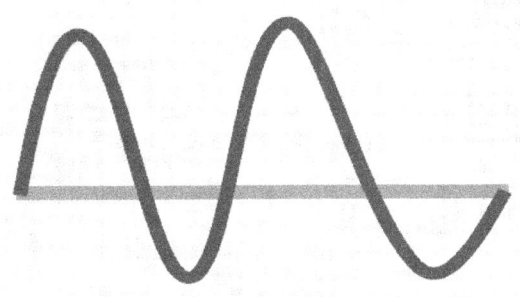

Modified Sine Wave

A modified sine wave power has a waveform more like a square wave, but with an extra step. Using a more basic form of technology than pure sine wave inverters, they produce power which is perfectly adequate for powering basic electronics. A modified sine wave inverter can be used for simple systems that don't have any delicate electronics or audio equipment that may pick up the choppy wave.

A modified sine wave inverter works fine with most equipment, although the efficiency or power of the material will reduce over some time. Use modified sine wave inverters to provide power for less sensitive appliances like phone chargers, heaters, and air conditioners. Appliances with electronic timers or digital clocks will often not operate correctly. Many appliances get their timing from the peak of the line power – basically, the modified sine has a flat top rather than a peak – this may cause the occasional double trigger. Because the modified sine wave is noisier and rougher than a pure sine wave, clocks, timers, radios and power drills may run faster or not work at all. . Modified sine wave inverters switch is DC supply between positive and negative poles to give you a simulated sine wave form. Use these inverters for televisions, laptops, digital microwaves, fridges, and other sensitive electronic equipment. Old tube TVs and motors with brushes are usually ok with modified sine wave A whole raft of modern appliances won't run as well and some not at all on this waveform: Laser printers, photocopiers, and anything with an electrical component called a thyristor,Anything with a silicon-controlled rectifier (SCR), like those used in some washing machine controls ,A few laptop computers, Some fluorescent lights with electronic ballasts, Some battery chargers for cordless tools, Some new furnaces and pellet heaters with microprocessor controls, Digital clocks with radio, Appliances having speed/microprocessor controls (like some sewing machines).

Square Wave

There are very few, but the cheapest inverters are square wave. A square wave inverter will run simple things like tools with universal motors without a problem, but not much else. Square wave inverters are seldom seen anymore. The square wave inverter are large obsolete as a result of the wave front shape that is not well-suited for running the modern appliances. It will cause greater damage in driving an inductive appliances e.g. Pumping machine.

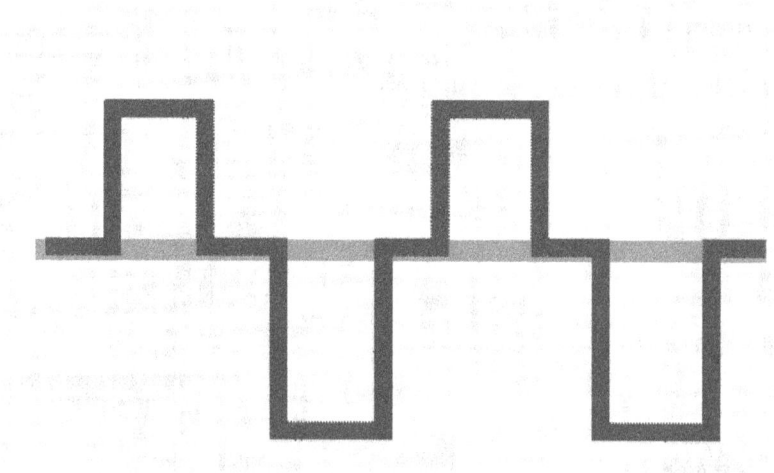

2.4 REVIEW OF INVERTER CAPACITY

2.4.1 SAFETY OF INVERTER

1. Safety When Operating Inverters are as follows;

- Installation and wiring compliance

-Installation and wiring must comply with the Local and National Electrical Codes and must be done by a certified electrician.

2. Prevention of electrical shock - Always connect the grounding connection on the unit to the appropriate grounding system.

- Disassembly / repair should be carried out by qualified personnel only.

- Disconnect all AC and DC side connections before working on any circuits associated with the unit. Turning the on/off Key on the unit to off position may not entirely remove dangerous voltages.

- Be careful when touching bare terminals of capacitors. The capacitors may retain high lethal voltages even after the power has been removed. Discharge the capacitors before working on the circuits.

3. Installation environment - The inverter should be installed indoor only in a well ventilated, cool, dry environment.

 - Do not expose to moisture, rain, snow or liquids of any type.

- To reduce the risk of overheating and fire, do not obstruct the suction and discharge openings of the cooling fans.

- To ensure proper ventilation, do not install in a low clearance compartment.

4. Preventing fire and explosion hazards.

- Working with the unit may produce arcs or sparks. Thus, the unit should not be used in areas where there are flammable materials or gases requiring ignition protected equipment.

These areas may include spaces containing gasoline-powered machinery, fuel tanks, and battery compartments.

5. Precautions when working with batteries

- Batteries contain corrosive diluted sulphuric acid as electrolyte. Precautions should be taken to prevent contact with skin, eyes or clothing.

- Batteries generate hydrogen and oxygen during charging resulting in evolution of explosive gas mixture. Care should be taken to ventilate the battery area and follow the battery manufacturer's recommendations.

- Never smoke or allow a spark or flame near the batteries.

- Use caution to reduce the risk of dropping a metal tool on the battery. It could spark or short circuit the battery or other electrical parts and could cause an explosion.

- Remove metal items like rings, bracelets and watches when working with batteries. The batteries can produce a short circuit current high enough to weld a ring or the like to metal and thus, cause a severe burn.

- If you need to remove a battery, always remove the ground terminal from the battery first. Make sure that all the accessories are off so that you do not cause a spark.

6. Preventing Paralleling of the AC Output

The AC output of the unit should never be connected directly to an Electrical Breaker Panel / Load Center which is also fed from the utility power / generator. Such a direct connection may result in parallel operation of the different power sources and AC power from the utility/generator will be fed back into the unit which will instantly damage the output section of the unit and may also pose a fire and safety hazard. If an Electrical Breaker Panel / Load Center is fed from this unit and this panel is also required to be fed from additional alternate AC sources, the AC power from all the AC sources like the utility / generator / this unit should first be fed to an Automatic / Manual Selector Switch and the output of the Selector Switch should be connected to the Electrical Breaker Panel / Load

Center. CAUTION: To prevent possibility of paralleling and severe damage to the unit, never use a simple jumper cable with a male plug on both ends to connect the AC output of the unit to a handy wall receptacle in the home / RV.

7. Preventing Input Over-Voltage

It is to be ensured that the DC input voltage of this unit does not exceed 16.8 + / - 0.3 VDC for the 12V battery version and 33.6 + / - 0.6 VDC for the 24V battery version to prevent permanent damage to the unit. Please observe the following precautions:

- Ensure that the maximum charging voltage of the external battery charger / alternator / solar charge controller does not exceed 16.8 - 0.3 VDC for the 12V battery version and 33.6 + / - 0.6 VDC for the 24V battery version.

- Always use a charge controller between the solar panel and the battery.

- Do not connect this unit to a battery system with a voltage higher than the rated battery input voltage of the unit (e.g. do not connect the 12V version of the unit to 24V battery system or the 24V version to the 48V Battery System).

8. Preventing Reverse Polarity on the Input Side

When making battery connections on the input side, make sure that the polarity of battery connections is correct (Connect the Positive of the battery to the Positive terminal of the unit and the Negative of the battery to the Negative terminal of the unit). If the input is connected in reverse polarity, DC fuse(s) inside the inverter will blow and may also cause permanent damage to the inverter.

9. Using Generator as External Input Source with Inverter Chargers

The AC output voltage of a generator is proportional to its rotational speed (RPM – Revolutions per Minute) and the current fed to its field windings. The frequency of the AC output voltage produced by the generator is proportional to the RPM of the engine and the number of poles used in the generator. The RPM of the generator is controlled and kept constant by the mechanical governor installed on the engine that is driving the generator. The output voltage of the generator is controlled by its electrical voltage regulator, which controls the current fed to its field windings. When an electrical load is applied to the generator, its output voltage tends to drop and the speed of the engine also tends to drop leading to drop.

-Do not put a power inverter near or on the radiator or heating vent. Do not put a power inverter directly under the sunlight. Ideal temperature for operation is from 50°F to 80° F.

-In order to effectively spread the heat produced when the power inverter is operating, you should allow fresh to enter. In addition, remember to keep enough space at the side and the top of the power inverter.

-Do not operate the power inverter close to the inflammable substances. Do not put the power inverter in places like battery section in which gas might amass.

CHAPTER THREE

METHODOLOGY
Design Requirements

The inverter operates by performing two functions. Firstly, it converts DC voltage

(12V) from the battery to AC voltage (12V) using a pair of powerful MOSFETs (Q1 and Q2)

serving as switches. Secondly, it steps up the resulting AC into equivalent mains voltage with

corresponding frequency and phase, using appropriate step up transformer. The positive 12V

from the battery is connected to the centre tap of the transformer primary, while each

MOSFET is connected between one end of the primary winding and ground. So by switching

on Q_1, the battery current can be made to flow through the top half of the transformer primary

and to ground through Q_1. Conversely, by switching on Q_2 instead, the current is made to

flow in the opposite direction via the lower half of the primary and to earth. Thus, by

switching on the two MOSFETs alternately, through an oscillator at a frequency of 50Hz and

AC voltage is applied across the primary windings, which induces an AC voltage in the

secondary windings. [13].

The DC-AC inverter consists of a battery which supplies the 12V DC input voltage to the

circuit. Then the PWM control circuit which is used to pulse the half bridge inverter. The half

bridge inverter will chop up the 12V DC supplied by a battery so that an AC is seen by the

transformer. The transformer is responsible for boosting the voltage by stepping up the

voltage[2].

The following equations were used to calculate the modulation amplitude and modulation

frequency for the PWM signal:

$Amplitude\ Modulation = V_{control} \div v_{tri}$

$Frequence\ Modulation\ Ratio = f_s \div f$

Where; Vcontrol is the peak amplitude of the reference sine wave with frequency off 1 Strike the peak amplitude of the saw-tooth wave with frequency offs.

3.1 DESCRIPTION OF MATERIALS FOR CONSTRUCTION

The followings are the component needed for the design and construction of 1KVA solar inverter;

- MOSFET

It main function is that it is used as a switch. The MOSFET used is IRF3205 which is a high current N-Channel MOSFET that can switch currents up to 110A and 55V.. It has segment;

-Source: Current flows out through Source.

-Gate: Controls the biasing of the MOSFET.

-Drain: Current flows in through Drain.

Fig.1: N-channel mosfet

- IC741

It is known as operational amplifier and it is a one kind of solid state IC. It consists of two inputs and two outputs which are inverting and non inverting terminals. This IC 741 is most commonly used in various electrical and electronic circuits. The main aim of using this IC741 is to strengthen AC & DC signals and it applications involves filters, comparators, pulse generators, oscillators, etc. IC 741 is made from various stages of transistor which commonly have three stages and also comprises of a set of FETs or BJTs.

30

The IC 741 operational amplifier has eight pins and their functions are;

Pin 1: is offset null

Pin 2: is inverting terminal.

Pin 3: is a non-inverting terminal.

Pin 4: is -Ve voltage supply (VCC).

Pin 5: is offset null.

Pin 6: is the output voltage.

Pin 7: is +ve voltage supply (+VCC).

Pin 8: is not connected; the most significant pins are 2, 3 and 6. Where pin 2 and 3 denote inverting and non inverting terminals and pin 6 denotes output voltage.

- IC 4N35

This is an optocoupler integrated circuit that is analogous to a relay which isolates two circuits magnetically. They differ with relays in the sense that they are smaller in size and allow fast operation. IC 4N35 is make-up of a led and a photo transistor and It has six pins.

- LM317

  This is a three pin adjustable regulator IC.

- IN4007

This is known as rectifier diode. This is a very simple common rectifier diode often used for reverse voltage protection that is it prevent current or voltage from been reversed.1N4007 is rated for up to 1A per 1000V.

- IN4148

This is called a signal diode which is a standard silicon switching signal diode. It is one of the most popular and long-lived switching diodes because of its dependable specifications.

- 10K Ohms

This resistor has brown, black, orange as it color code with Gold has it last stripe to represent tolerance. Gold means ±5% tolerance. It is used to limit the flow of current.

- 5.6K Ohms

This is also called a fixed resistor. It helps to reduce the flow of electricity in a circuit with carbon-film resistor with ±5% tolerance.

- 47K Ohms

This resistor is a fixed resistor which is also used for limiting current with a 0.4Watt and 5% tolerance.

- LM339

It is a comparator IC with four inbuilt comparators. A comparator is a simple Circuit that moves signals between the analog and digital worlds. It is used for high Voltage cutoff, low voltage cutoff and battery overcharged cutoff. It has four pins;

Pin1: it is the second output pin.
Pin2: it is the first output pin.
Pin3: it is the positive power supply pin.
Pin4: it is the inverting input pin.

- VARIABLE RESISTOR

A variable resistor is a resistor of which the electric resistance value can be adjusted. It is used to increase the frequency range.

Fig2: Variable resistor

- WIRE

32

It is used to transfer the flow of voltage and current.

- TRANSFORMER

This is a device that works by the principle of mutual induction. It used to step up or step-down voltage supply.

- RELAY

A relay is a mechanical switch used to switch other circuit on and off. A relay
Will have normal operating voltage in which it coil operate and maximum current rating.

3.2 DESIGN ANDCONSTRUCTION.

Inverter Circuit

The circuit diagram for this project consist of The Oscillator, The Driver and the Output Amplifier, Inverter Transformer, The battery Charger, The Change Over, Inverter AC Output, Protections , Indicators and switch and it is shown below;

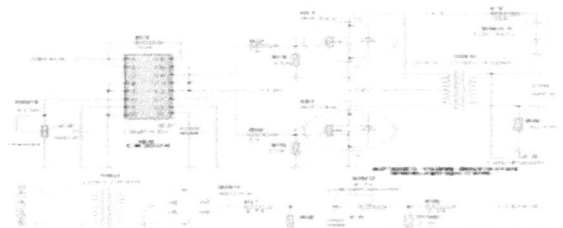

Fig3: circuit diagram of inverter [3].

The following lists are the sections that make up an inverter;

1. Oscillator (Frequency) Unit: An oscillator can be though to fasten amplifier which generates or provides itself with an input signal and an output frequency which is determined by the characteristic of the selected device. To cause the oscillation to be self driven, the signal feedback must be regenerated (positive feedback).The oscillator used in this work is IC SG3524.Different oscillator circuit can be used but we adopt SG3524oscillator circuit in this

33

project which has provision for output feedback. This feedback allows us to generate a constant output voltage of 220V no matter the battery voltage. This section uses a pulse width modulator (PMW) IC SG3524 to generate the 50Hz frequency required to generate AC supply by the inverter.

The frequency of oscillation of the IC SG3524is programmed by RT and CT of the oscillator according to the approximation formula:

$f = 1.18 \times RC \times CT$

Where; RC is the timing resistor in kΩ (ranges from 1.8KΩ to 100KΩ).

CT is the timing capacitor in µF (ranges from 0.001µF to 0.1µF).

Hence, 23KΩ was selected through a variable resistor and a Capacitor of 0.1µF was used.   Thus frequency is given by;

$f = 1.18 \times 23000 \times 0.000001 = 50Hz$

Thus, the switching period T is;

$T = 1 \div f = 1 \div 50 = 0.02s$

2. MOSFET assembly unit: This unit consists of an array of MOSFETs. The MOSFET used in this construction is the depletion mode MOSFET with path number IRFP150N having a current rating of 39A, voltage rating of 100V and power rating of 190W.

The required number of MOSFETs per channel as determined for 1KVA solar inverter is given by;

Number of MOSFETs = 1000 ÷ 190 = 5.3

Thus, approximately 5 MOSFETs are required per switching channel to be connected in parallel to boost the current to drive the transformer.

3. Transformer unit: The transformer used for this project has a center-tapping which divides the primary into two equal sections. This center-tapping is connected to the positive terminal of the battery. Two ends of the primary are connected to the negative terminal of the battery through switches S1 and S2. These switches S1 and S2 are turned ON/OFF alternatively to generate current in the primary coil. When the switch S1 is closed and S2 is opened, the current flows in the first part of the primary winding and the EMF is induced in the secondary winding. When the switch S2 is closed and S1 is opened, the current flows in the second part of the primary winding and the EMF of opposite polarity is induced in the secondary winding .Thus, if the switches S1 and S2 are alternatively opened and closed at constant rate, then the output from the secondary winding is a square wave of the frequency at which the switches S1 andS2 are opened and closed. A single transformer was used for this construction.

The number of secondary windings is obtained from the approximate design equation for 50Hz transformer using laminated E-core as:

Secondary turns = 48xOutput voltage core area leading to a calculated value of 429turns.The primary winding was obtained from the voltage transformation ratio of a transformer as follows:

$$\frac{E2}{E1} = \frac{N2}{N1}$$

Hence, $N_1 = \frac{E1N2}{E2} = {}^{12 \times 429}/_{220} = 23.4\text{turn}$

35

4. Battery charger monitor unit: This unit is responsible for cutting off or regulating the charging of the battery to prevent excessive charging of the battery. It consists of transistors, resistors, diode, zener-diode and a 12V relay device.

5. The Driver and the output amplifier: The MOS drive signals are given to the base of MOS driver transistor which results in the MOS drive signal getting separated into two different channels. The transistors amplify the 50Hz MOS drive signal at their base to a sufficient level and output them from the emitter. The50Hz signal from the emitter of each of the transistor is connected to the gate G of all the MOSFETs in each of the MOSFET channel, through the appropriate resistance.

6. The battery supply: This is connected to the IC S3524 through the inverter ON/OFF switch. The flip-flop converts the incoming signal into signals with changing polarity such that in a two signal with changing polarity with which the first is positive while the second is negative and vice versa. This process is repeated 50times per second to give an alternating signal with 50Hz frequency at the output of SG3524.This alternating signal is known as "MOS Drive Signal".

7. The battery Charger: When the inverting circuit senses supply from the panel, it stops operation but the charger section in the inverter starts its operation. In this mode, the inverter transformer works as a step down transformer and output 12V at its secondary winding. During the charging, MOSFET transistors at the output section works as rectifier with the drain working as the cathode while the source works as the anode. The center-tapping of the transformer receives positive supply and the MOSFET source 'S' receives negative supply from the battery. The center-tapping is connected to the positive terminal of the battery and the MOSFET source S is connected to the negative terminal with a shunt resistance.

8. The change over: This section is used to switch ON the inverter when there is supply from the solar panel and to switch OFF the inverter when supply returns. During changeover, when

the inverter receives solar supply, it stops drawing the battery supply and the AC mains supply inverted by the inverter is directly sent to the inverter output socket. This is done using a one, two-pole relay.

9. Inverter AC Output: The AC output gives a 220V AC, 50Hz either directly from the input when the DC voltages are supply from the panel or from the inverter circuit action on the battery. Computers and other household appliances are connected to this output.

10. Switch: A switch is connected to the front of the inverter. It is used to close or open the inverter circuit.

11. Charge controller: This inverter has an inbuilt controller that regulates the battery from being either overcharged by the solar panels or use below the manufactures specification. The controller will be able to handle a maximum current of 20A and a maximum voltage of 50V. The charged controller is designed to work for both 12V and 24V panel(s).

12. IC circuit: The IC circuit has 16pins and each pin uses are as follows;

Pin1: This is provided with error sample voltage from the feedback network.

Pin3 and 9: These pins are not used.

Pin4 and 5: These pins are connected to ground.

Pin6: Timing resistor is collected to pin6 and ground tp generate 50Hz frequency of oscillations of the IC.

Pin7: Timing capacitor is connected to pin7.

Pin8: This is the ground pin.Pin15: It is the power pin connected to the 12Vpowersupply from the battery.

Pin10: This is the shutdown pin.

Pin11 and 14: These are the oscillator section of the oscillator IC and are also the Output pins

37

which generate 4.5V.

Pin12 and 13: They deal with battery voltage.

Pin16: It is the regulator pin which gives out 5Voutput.

3.3 CONSTRUCTION

The construction of this project started with the building of the transformer from the laminating core, followed by the rectification stage, sensing and monitoring stage, comparator and transistor switching.

3.3.1 TOOLS FOR CONSTRUCTION

The following are the tools used for design and construction;
1. METER: It is used for measuring and testing electronics component.
2. SCREEW DRIVER: It is used for driving screws.
3. SPANERS: It is used for loosing and tightening bolts

4. DRILLING MACHINE: It is used for drilling holes in metals.
5. SOLDERING IRON AND LEDS: It is used for soldering components on circuit boards.

3.4 ESTIMATION OF LOADS TO BE POWERED BY THIS PROJECT;

The table below shows the load estimates of various appliances and their power ratings

Appliances	Quantity	Power rating
1.Desktop computer	1	360 W
2.Lighting point	2	60 W
3. Fridge	1	105 W

Analysis of Power rating

Inverter power output (P) = 1000 Watts

Output voltage, V = 220 V

Inverter Input = battery output voltage =12 V

Frequency = 50 Hz

Power factor = 0.8

Real power of inverter = Apparent power × power factor

$$= 1000 × 0.8 = 800VA$$

Therefore, the load current flowing at the transformer primary

;   Real power = current (I) × voltage

$$1000 = I × 12$$

Hence,   I = 1000 ÷ 12 = 83.3A

3.5 The Casing

Figure a: front view of inverter casing   Figure b: back view of inverter casing

Packaging of the constructed project was done to achieve a good looking and presentable device. During the packaging, some factors were considered; these include:

1. The durability of the material to be used in the packaging, materials like wood, plastic or metal could be used but for this project work, metal sheet was used; this is to ensure easy dissipation of heat to the environment.

2. Again, caution was taken to avoid short-circuiting of any part of the design. The portability of the package was taken into consideration to limit the space it will occupy as well as to ease the burden associated with the movement of the device.

3. The ventilation of the package was also considered; this is to help in temperature control of the device since most of the components in the construction are heat-generating components.

CHAPTER FOUR

RESULT ANALYSIS AND DISCUSSIONS

Experimental Set-Up

Components were connected as shown in the circuit below;

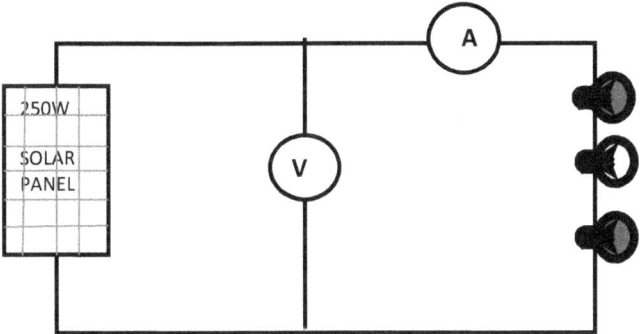

Figure1: circuit model for the experiment

The figure above shows the setup of the experiment performed. From the experiment

setup the ammeter is connected in series while the voltmeter is parallels the mono-crystalline

solar panel. Also the three bulbs are connected in series with each other to the solar panel

because the voltage rating of each bulb is 12 V. The total voltage rating will be 36 V which is

equivalent to the voltage rating of the solar panel.

Specifications table of the solar panel used for experiment.

Table 1;

Type	Mono crystalline
Maximum power (Pmax)	250 W
Open circuit voltage (Voc)	37.8 V
Short circuit current (Isc)	8.8A
Max power voltage (Vmp)	30.5 V
Max power current (Imp)	8.22A
Max system voltage (Voc)	1000V
Power tolerance	±3%

From the specification table above, the followings are calculated;

i. Maximum power (Pmax) can be calculated as;

$$P_{max} = I_{max} \times V_{max}$$

And maximum power can also be calculated from s the short circuit current I$_{sc}$ and open circuit voltage Voc as;

$$P_{max} = FF(I_{sc} \times V_{oc})$$

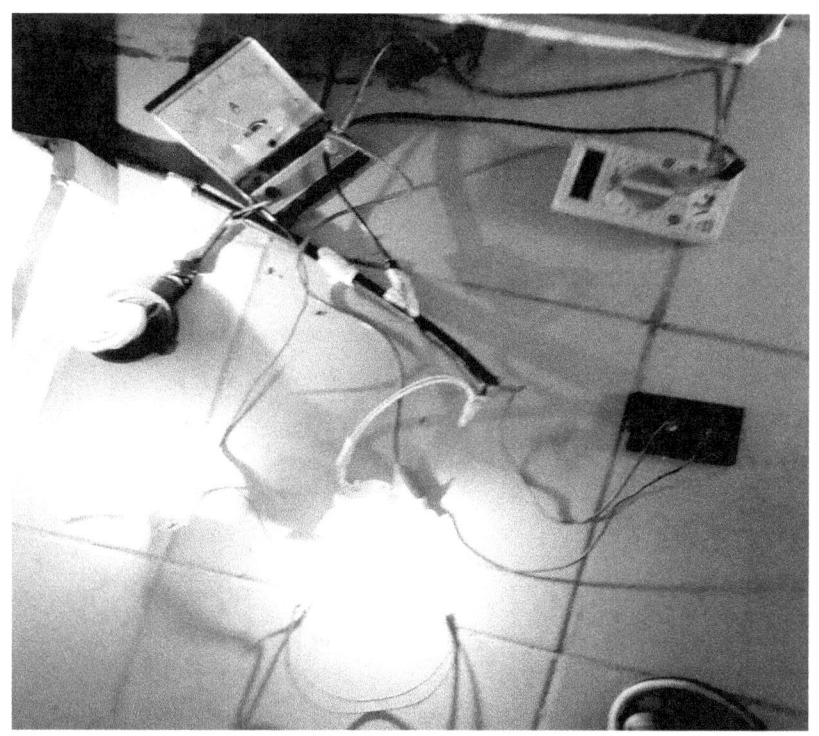

Figure2: setup when circuit is closed.

  ii. Fill factor

$$fill\ factor = \frac{Area\ A}{Area\ B}$$

OR

$$\frac{maximum\ power}{Voc \times Isc}$$

Maximum power is 77.98

Voc is 37.8

Isc is 8.8

$$\frac{maximum\ power}{Voc \times Isc}$$

$$= \frac{77.98}{332.64}$$

43

=0.2344

So the fill factor is 0.2344

4.1 RESULT AND ANALYSIS

The parameters such as current, voltage and power output were calculated for a mono-crystalline solar panel and also this circuit was operated for a day. The table2 shows the readings recorded under different atmospheric conditions;

Table 2; Variation of Voltage, Current and Powered under different Atmospherics condition in a Day

S/N	Time/ min.	Period/ sec	Voltage/ V	Current/ A	Weather	Power/ W
01.	09:30am	30	10.70	00.60	Rainy	06.42
02.	10:00am	60	31.80	02.10	Sunny	66.78
03.	10:30am	90	18.70	00.90	Rainy	16.83
04.	11:00am	120	32.10	02.20	Sunny	70.62
05.	11:30am	150	33.50	02.20	Sunny	73.70
06.	12:00pm	180	26.50	02.10	Cloudy	55.65
07.	12:30pm	240	08.90	00.60	Rainy	05.34
08.	01:00pm	300	32.10	02.20	Sunny	70.62
09.	01:30pm	360	08.20	01.20	Rainy	09.84
10.	02:00pm	420	32.30	02.40	Sunny	77.52
11.	02:30pm	480	33.00	02.40	Sunny	79.20
12.	03:00pm	540	32.20	02.40	Sunny	77.28
13.	03:30pm	600	33.20	02.40	Sunny	79.68
14.	04:00pm	660	30.30	02.20	Sunny	66.66
15.	04:30pm	720	14.50	01.70	Rainy	24.65

16.	05:00pm	780	06.60	01.20	Rainy	07.92
17.	05:30pm	860	00.90	00.50	Cloudy	00.45
18.	06:00pm	940	00.20	00.32	Cloudy	0.064

The table above analyzed the result from the experiment performed. It consists of the following parameters; the voltage, the current, the power, the period and the time under different climatic atmosphere condition

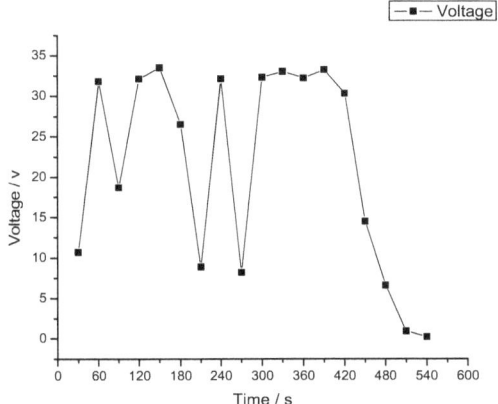

Figure 3: Variation Graph of Voltage against Time

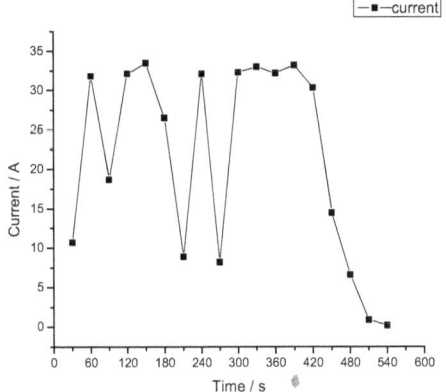

Figure 4: variation graph of Current against Time.

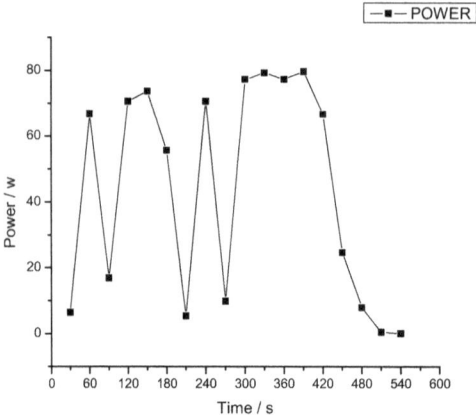

Figure 5: Variation Graph of Power against Time.

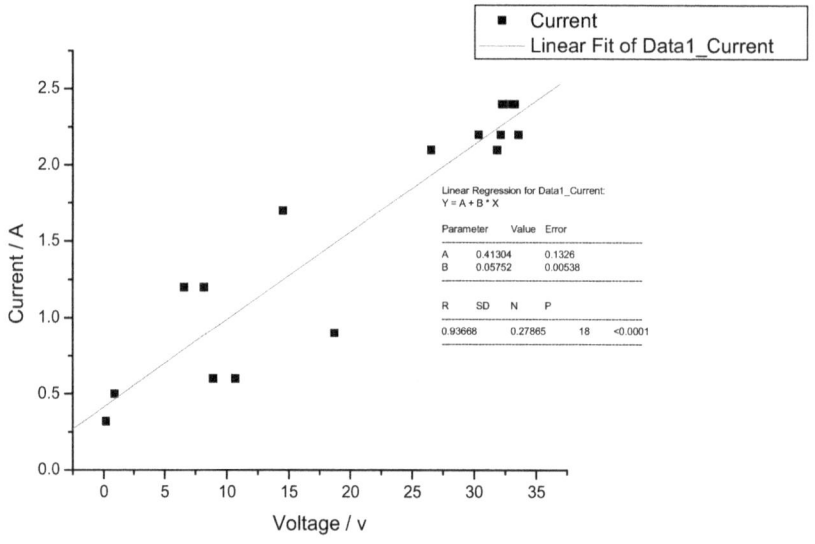

Figure6: Correlation Graph of Current against Voltage

46

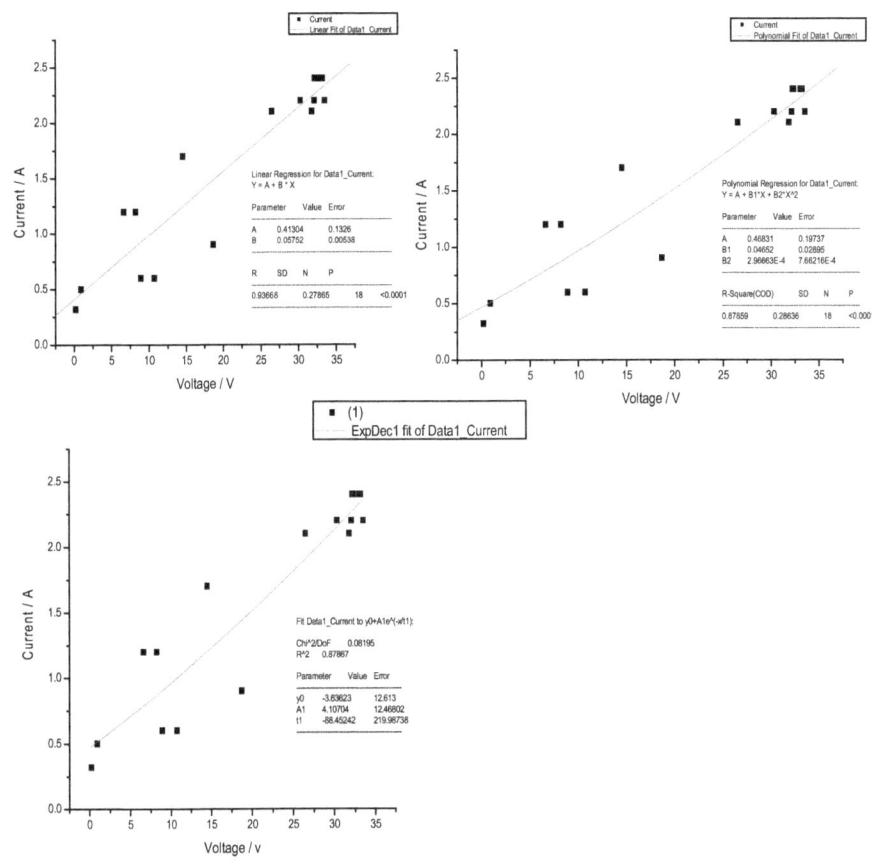

Figure 7: Correlation Graph of Current against Voltage

4.2 DISCUSSION

Figures 3; shows the variation plots of voltage against time for a mono-crystalline panel under different atmospheric conditions. The horizontal axis (Voltage) is dependent variable and the vertical axis (Time) is independent variable. The minimum and maximum values of the voltage are 00.20V and 33.50V occurring at 9: 30 am and 11:30 pm respectively.

It was observed that when the sky was bright the voltage increases and when it was raining the voltage decreases. Also the voltage finally drops to it starting point when the sky becomes dark at night.

47

Figure 4; shows the variation plot of current against time for a mono-crystalline panel under different atmospheric conditions. The horizontal axis (Time) is independent variable and the vertical axis (current) is dependent variable. The minimum and maximum values of the current are 00.32A and 02.40A occurring at 9: 30 am and 11:30 pm respectively. Also when the sky is bright, the current increases and when it is raining the current decreases. Also the current finally drops to it starting point when the sky becomes dark at night.

Figure 5; shows the variation plot of power against time for a mono-crystalline panel under different atmospheric conditions. The horizontal axis (Power) is dependent variable and the vertical axis (Time) is independent variable. The minimum and maximum values of the current are 0.064W and 79.68W occurring at 9: 30 am and 11:30 pm respectively. It was observed that when the sky was bright the power increases, when it was raining the power decreases. Also the power finally drops to it starting point when the sky becomes dark at night.

Figure6; shows I-V relationship of the panel. Which are the linear graph of current against voltage for a Mono-crystalline panel and the gradient of the graph gives the conductivity of the solar panel. R = 0.93668

Equation for linear regression is $Y = A + B * X$

Mathematically,

$V = IR$

This implies that; $\frac{I}{V} = \frac{1}{R} = Condutivity$

Figure 7; show the correlation graph of current against voltage using a linear fit, exponential fit and polynomial fit. The gradient for each cases of the graph gives the conductivity of the mono-crystalline solar panel under different atmospheric conditions. It was observed that the graph of current against voltage for various fit has a very strong positive correlation. The correlation value for polynomial is $R^2 = 0.87859$.

Polynomial fit $Y = A + B_1 * X_1 + B_2 * X_2$

The Equation for linear fit is $Y = A + B * X$ and the value of R = 0.93668

For exponential fit the value of R = 0.87867.

CHAPTER FIVE

CONCLUSION AND RECOMMENDATION

CONCLUSION

This write up entails the details of the design, construction and analysis of 1KVA inverter powered by solar panel via a 12V industrial battery. The system is tested under different climatic conditions of Osun State University, Oshogbo. It was observed in the results obtained that there were a strong variation between the current, the voltage and power of the mono-crystalline solar panel under different weather condition. It was observed that any rise in the voltage of the solar panel will lead to rise in both the power and current of the solar panel which leads to rise in the state of charge of the storage battery.

As the solar radiation increased, voltage of PV module has also increases which cause a direct proportion increase in photovoltaic module electrical current and power. So it can be concluded that the performance of solar power inverter does not only depends on voltage but also on current and power of the solar panel.

This project has met the objective and purpose for which it was designed and constructed. It could be used in homes, offices and industries to serve as an alternative power supply because of the following advantages:-

- Low maintenance cost

- No moving parts

- No noise pollution

- Easy installation.

- No environmental pollution.

5.1 RECOMMENDATION

One of the limitations of this project is that the 12V DC battery backup cannot withstand a large load applied on it for a long period; the overload cut-off is triggered when the load on the Inverter is larger than the designed capacity. Hence, for a more reliable and stable power supply, it is recommended that a larger battery backup should be provided to enable this project withstand larger loads for a long period. The department should develop a cottage workshop where young interested students will be engaged in mass-production, customization and commercialization of this equipment (Inverter); thereby encouraging entrepreneurship and product localization.

Therefore, I hereby implore the Federal Government of Nigeria to focus more on the grid expansion of green energy as this will increase the rate at which electricity is being generated.

REFERENCES

1. Ellabban, Omar; AbuRub, Haitham; Blaabjerg, Frede (2014)."Renewableenergyresources: Current status, future prospects and their enabling technology".(Publisher)

2. REN21 (2010).Renewables2010GlobalStatusReport

3. World Energy Assessment (2001).Renewable energy technologies *(Publisher)*

4. Keyword World Energy Statistics (2018).International energy agency

5. REN21, Global status report (2016)

6.Armaroli,Nicola;Balzani,Vincenzo(2016).Solarelectricityandsolarfuel:Statusandperspective inthecontextoftheenergytransition.

7. IEA Renewable Energy Working party (2002).

8.Jacobson,MarkZ;etal.,(2015).:100%cleanandrenewablewind,water,andsunlight(WWS)allsectorenerg yroadmapforth50UnitedStates.

9. Design construction and performance evaluation of 1Kva pure sine wave power inverter. Sheu Akeem Lawal and Alade Olusope Michael (2015)

10. International Journal of Scientific & Engineering Research, Volume6, Issue3, March-20151292ISSN2229-551.

11. Design and implementation of a 5kvA inverter (2016). Chukwuka Anene

Contents

Publisher: Eliva Press SRL

Email: info@elivapress.com

Eliva Press is an independent publishing house established for the publication and dissemination of academic works all over the world. Company provides high quality and professional service for all of our authors.

Our Services:
Free of charge, open-minded, eco-friendly, innovational.

-All services are free of charge for you as our author (manuscript review, step-by-step book preparation, publication, distribution, and marketing).
-No financial risk. The author is not obliged to pay any hidden fees for publication.
-Editors. Dedicated editors will assist step by step through the projects.
-Money paid to the author for every book sold. Up to 50% royalties guaranteed.
-ISBN (International Standard Book Number). We assign a unique ISBN to every Eliva Press book.
-Digital archive storage. Books will be available online for a long time. We don't need to have a stock of our titles. No unsold copies. Eliva Press uses environment friendly print on demand technology that limits the needs of publishing business. We care about environment and share these principles with our customers.
-Cover design. Cover art is designed by a professional designer.
-Worldwide distribution. We continue expanding our distribution channels to make sure that all readers have access to our books.